Process Intensification 2018

Topical Conference at the 2018 AIChE Spring Meeting and
14th Global Congress on Process Safety

Orlando, Florida, USA
22 – 25 April 2018

ISBN: 978-1-5108-6435-1

Printed from e-media with permission by:

Curran Associates, Inc.
57 Morehouse Lane
Red Hook, NY 12571

Some format issues inherent in the e-media version may also appear in this print version.

Printed by Curran Associates, Inc. (2018)

For permission requests, please contact AIChE
at the address below.

AIChE
120 Wall Street, FL 23
New York, NY 10005-4020

Phone: (800) 242-4363
Fax: (203) 775-5177

www.aiche.org

Additional copies of this publication are available from:

Curran Associates, Inc.
57 Morehouse Lane
Red Hook, NY 12571 USA
Phone: 845-758-0400
Fax: 845-758-2633
Email: curran@proceedings.com
Web: www.proceedings.com

TABLE OF CONTENTS

(68b) Modular Systems for Clean Coal 1
Bhima Sastri

(68c) NSF Funding Opportunities for Fundamental Research on Modular Chemical Process Systems 2
T. J. Mountziaris

(68d) Chemical and Thermal Process Intensification at ARPA-E 3
Addison K. Stark

(93a) Industrial Application of Process Intensification in Adsorptive Systems 4
Yang Luo

(93b) Enabling Modular Process Intensification through Microfibrous Entrapment Technology 5
Paul S. Dimick, Hongyun Yang, Bruce J. Tatarchuk

(106a) cMIST Compact Inline Gas Treating Technology 6
Shwetha Ramkumar, Norman K. Yeh, Don P. Shatto

(106b) Novel Cryogenic CO_2 Removal- Operating Conditions and Economic Analysis 7
Farhad Fazlollahi

(106c) Dynamic Intensification of Separation Processes 8
Lingqing Yan, Thomas F. Edgar, Michael Baldea

(106d) Top Dividing Wall Columns in Process Intensification 9
Manish Bhargava, Roomi Kalita, Brad Fleming

(131a) Use of Dividing Wall Technology to Increase Throughput and Improve Product Specifications in Two-Diameter Columns 15
Manish Bhargava, Roomi Kalita

(131b) Process Intensification of Distillation Using a Microwick Technology Platform 16
Ward E. Tegrotenhuis, Danny Bottenus, Paul Humble, Michael R. Powell

(131c) Improving Plant Economics By Upgrading Low Value Hydrocarbon Stream By Enrichment of C_{9+} Compounds Using Divided Wall Distillation Columns 17
Abdullah Alqahtani, Mahesh Kumar Srinivas

(163a) Intensified-By-Design (IbD): Creating a Platform for Facilitating Process Intensification in Solids Handling Applications 18
Kamelia Boodhoo, Anh Phan, Vladimir Zivkovic, Valentine C. Eze, Ahmad Mustaffar

(163b) Rapid, In-line Characterization to Support Process Intensification Concepts 19
Tyler L. Westover

(163c) Hybrid Heat Pipe Screw Dryer: A Novel, Continuous and Highly Energy-Efficient Drying Technology 20
Ahmad Mustaffar, Kamelia Boodhoo, Anh Phan

(163d) Process Intensification and Cost Reduction for in-Plant Biomass Preprocessing Equipment 21
Michael J. Blaise, Nicholas R. Schwartz, Paul E. Yelvington

(173c) Overcoming Heat Transfer Bottlenecks in Non-Equilibrium Chemical Processes through Autothermal Operation 22
Robert C. Brown, Chad Peterson

(173d) Reverse Flow Reactors: Process Intensification for High Temperature Chemistries 23
Federico Barrai, Dhaval Bhandari, Changmin Chun, Frank Hershkowitz, James Lattner, Elizabeth G. Mahoney, Brian M. Weiss

(190a) Building with Blocks 24
Greg Theunick

(190c) Critical Factors and Criteria for the Development and Deployment of Modular Chemical Processes 25
Paul S. Dimick, Bruce Tatarchuk

(190d) Continuous Processing of Specialty Chemicals Opportunities with MCPI 26
Oleg Pajalic

Author Index

Some abstracts in this book refer to images that were not made available by the publisher

512577 Modular Systems for Clean Coal

Abstract:

There Are Many Technical and Manufacturing Challenges That Face Our Collective Efforts to Build an Affordable, Clean, and Secure Energy Future. To This End, the U.S. Department of Energy (USDOE) Has Funded Research and Development of Technologies Focused on Assessing Strategies for Rapid Deployment of Modular Systems That Will Expand Market Penetration of Clean Energy Technologies at All Scales. Â Currently Usdoe Is Assessing the Potential for Modular Systems to Accelerate Technology Deployment By Leveraging Advance Manufacturing Techniques That Make Such Systems Competitive. This Report Will Specifically Aim to Discuss Interchangeable, Interoperable, and Robust Modular System Platforms That Separate Gases (CO_2, N_2, O_2), Convert Domestic Carbon Resources (coal, natural gas, biomass, and biogas) to Produce Value-Added Products (power, fuels, and chemicals) While Addressing Environmental Concerns (water, carbon, and other emissions). Specifically the Report Will Identify the Major Challenges Facing Modularization and Examine Possible Technical and Economic Solutions through an Integrated Systems Perspective. the Report Also Discusses Opportunities to Harness Innovations in Advanced Manufacturing (e.g. additive manufacturing, microfluidics, advanced simulation) to Reduce Costs, Improve Performance, and Consider Design Elements to Improve Deployment. The Author Will Discuss the Challenges for Making These Modular Systems Come to Marketplace

517173 NSF Funding Opportunities for Fundamental Research on Modular Chemical Process Systems

Abstract:

This presentation will focus on NSFâ€™s vision and funding opportunities for fundamental research related to Modular Chemical Process Systems (MCPS). MCPS have emerged as a potentially transformative concept, powered by the implementation of principles of Process Intensification (PI), that can yield substantial economic and environmental benefits for certain applications, when compared to traditional chemical plants. For example, MCPS can enable rapid deployment of new chemical and energy technologies, take advantage of geographically scattered feedstocks, and provide autonomous operation. Additional potential benefits of MCPS include reduced capital cost, improved handling of variations in feedstock supply rate and composition, reduced plant down-time, higher energy efficiency, increased safety, and waste minimization. Successful implementation of this new chemical manufacturing paradigm at commercial scales depends on the development of novel conceptual designs based on PI, a fundamental understanding of the underlying complex chemical and physical processes taking place in intensified process units, the development of methods for cost-effective manufacturing of modules, and the development of algorithms for real-time optimization and control of MCPS. This is an opportunity for convergent fundamental research using principles from process systems engineering, catalysis, reaction engineering, molecular thermodynamics, separations, transport phenomena, data science, and smart/additive manufacturing.

517213 Chemical and Thermal Process Intensification at ARPA-E

Abstract:

The US Department of Energyâ€™s Advanced Research Projectâ€™s Agency â€“ Energy (ARPA-E) has been active in funding projects focused on the intensification of chemical and thermal processes for improvements in industrial energy efficiency. In this talk Fellow and Acting Program Director Addison K Stark will highlight major areas of ARPA-E investment and discuss ongoing interests in process intensification.

513924 Industrial Application of Process Intensification in Adsorptive Systems

Abstract:

Process intensification is commonly defined as " any chemical engineering development that leads to a substantially smaller, cleaner, safer, and more energy efficient technology is process intensification"1 or "to create new processes through: reduced size, use of extreme forces or unusual contacting phases, process combinations and reconfigurations"2. Stankiwicz and Moulijn3 proposed to define the field into two areas: process intensifying equipment and process intensifying methods. Successful industrial application of process intensification requires thorough understanding of both process intensifying equipment and process intensifying methods.

Industrial application of process intensification for adsorptive system often focused on smaller system, short cycle time, and fast rate adsorbent.

To design an intensified pressure swing adsorption system, ideally, we will need fast adsorbent material with high capacity and selectivity, high efficiency equipment, reliable control system, accurate process model to size the system and predict the performance.

However, many times, equipment and design tools used for the conventional system are not suitable or adequate for the intensified system.

The presentation will cover the development of pressure swing adsorption system for industrial gas separation. The improvements on process modeling in mass transfer and kinetic constant for fast cycles, selection and balancing of equipment capability, process cycle, and adsorbent materials chacracterics will be discussed.

Examples of industrial application will be shared to illustrate the need of improved process intensifying equipment and process intensifying methods to achieve the intensified pressure swing adsorption system.

References:

1. Reay, Ramshaw, Harvey, Process Intensification, Oxford: Butterworth-Heinemann, p.21 (2008)

2. P.R. Westmoreland, Workshop on Process Intensification, Sept 30-Oct 1, (2014)

3. A.J. Stankiewicz, J.A.Moulijn, "Process Intensification: Transforming Chemical Engineering", Chemical Engineering Progress, 96, pg 22-34, (2000)

513239 Enabling Modular Process Intensification through Microfibrous Entrapment Technology

Abstract:

Fundamental rate phenomena and their interactions commonly limit the feasibility of modular chemical process intensification. Process intensification can be achieved using four distinct approaches which include integration of unit operations, integration of functions, integration of phenomena, and targeted enhancement of phenomena in operations [Eden and Gani]. When implementing any of these approaches, situations often arise when the heat and mass transport limitations of traditional designs and materials do not accommodate the fundamental transport needs necessary to properly intensify the system. When this occurs, engineers often attempt to employ methodologies that contain complex reactor structures and processing approaches to enhance these phenomena. These approaches typically never pass the bench or pilot scale due to limitations in reliability, manufacturability, and cost. It is necessary to simultaneously overcome these limitations to produce a modular process intensified system. For modular systems, it is often necessary to achieve economic improvements by implementing simplified systems with high manufacturability that enable cost reductions by numbering-up with assembly-line production approaches as opposed to custom field constructed systems.

IntraMicronâ€™s microfibrous entrapment technology is ideally suited to assist with process intensifying reactive systems that have fundamental heat and mass transport limitations. Microfibrous entrapment consists of entrapping active grains of catalyst and sorbent within a sinter-locked fibrous network. When this network is composed of metal fibers, the structure has an effective radial thermal conductivity that is 250-times greater than a traditional packed bed reactor system coupled with a 10-fold enhancement in mass transport phenomena. Simultaneously providing a fundamental enhancement in heat and mass transport in reactive systems greatly facilitates the other aforementioned process intensification methodologies including integrating functions, phenomena, and unit operations. IntraMicron and its partners have been investigating the application of this technology to systems that enable up to 9 levels of process intensification in a single vessel by combining numerous reaction steps and eliminating intermediate separations. The heat and mass transport benefits provided by microfibrous entrapment are akin to those provided by a fluidized bed reactor; however, microfibrous entrapment enables these benefits to be achieved in a simple, robust, scalable, easily-manufactured fixed bed system. For example, the application of this approach to Fischer-Tropsch Synthesis reactor enables up to a 4-fold enhancement in tube diameter, 16-fold reduction in tube count, and 4-fold reduction in overall reactor volume while providing enhanced reactor temperature uniformity and selectivity. Microfibrous materials are highly manufacturable because they are produced by robust, roll-to-roll wetlay and sintering processes. Furthermore, sintered microfibrous materials can be manufactured in a pre-form to enable the entrapment of catalyst and sorbent particles produced by any manufacturer. The enhanced performance, robustness, and manufacturability of microfibrous materials coupled with the simplicity of the reactors necessary to implement this technology make it ideal for implementation in modular process intensified systems.

This presentation will discuss several current and ongoing applications of microfibrous materials and their process intensification benefits including gas-to-liquids(GTL), reforming, ion exchange, pressure swing adsorption, and fine chemicals synthesis. IntraMicron has demonstrated several GTL reactors across the country at scales up-to 2 bpd and is in discussions to demonstrate this technology at scales of 25 â€“ 100 bbl/day for distributed energy applications. IntraMicron recently received a grant from RAPID to investigate the application of microfibrous entrapment technology to assist with the process intensification and modularization of Ion exchange systems for nuclear waste treatment and CO_2 removal from natural gas through pressure swing adsorption. The application of this technology for transitioning from batch to flow processing for fine/specialty chemicals will also be highlighted.

514266 cMIST Compact Inline Gas Treating Technology

Abstract:

cMIST™: Compact Inline Gas Treating Technology

Shwetha Ramkumar, Norman K. Yeh, Don P. Shatto, Chuck J. Mart

ExxonMobil Upstream Research Company

Background

The natural gas industry is increasingly pursuing compact and low weight processing technology to meet the technological and economic demands of offshore, onshore, remote, and challenged gas processing. ExxonMobil Upstream Research Company has developed the Compact Mass transfer and Inline Separation Technology (cMIST™) gas treating system, a novel, compact and low weight processing technology platform, to achieve process intensification in processing facilities.

Aims and Methods

The cMIST™ system can replace conventional absorbers with in-line equipment for mass transfer and separation and can be applied to gas dehydration, selective H_2S removal and low level CO_2 removal. The first application of the technology, which is dehydration of natural gas to pipeline specification, has been qualified and licensed to Sulzer Chemtech for deployment on offshore and onshore facilities. It was demonstrated using a skid-mounted test unit in the Woodford trend area of Oklahoma, USA at an XTO production facility.

The selective H2S removal application is currently in demonstration phase at an XTO gas plant in Teague, TX.

Results and Conclusions

The cMIST™ dehydration system was tested to examine the dehydration capacity, operability, and reliability and it achieved pipeline specification over a range of process conditions and during sustained operation. The cMIST™ system for selective H_2S removal is currently being demonstrated with MDEA solvent and is achieving pipeline specification with significantly lower CO_2 pickup compared to a conventional contactor over a range of process conditions.

The cMIST™ system enables 50% system weight reduction and 50 - 80% absorber footprint reduction along with significant capital cost savings. The cMIST selective H2S removal system has added benefits of lower solvent circulation rate and the potential for elimination of the Acid Gas Enrichment (AGE) unit. The technology is highly modular to allow for simple transportation into remote, challenged, or offshore environments and provide installation configuration flexibility. The technology is also beneficial for debottlenecking and capacity enhancement applications.

513675 Novel Cryogenic CO_2 Removal- Operating Conditions and Economic Analysis

Abstract:

Natural Gas (NG) pretreatment systems have been simulated to remove more than 99% of the CO_2 from NG. Steady-state simulations with Aspen Plus^{TM} illustrate the effects of different parameters on CO_2 capture percentage. Vapor mole fractions of CO_2 and methane also have been investigated by changing the pressure at constant temperature. Increasing pressure increases CO_2 capture and hence methane mole fraction in the vapor. Consequently, mole fraction of CO_2 in vapor decreases. As pressure increases, the methane in the vapor also condenses, creating more liquid in which solid CO_2 can dissolve. Increasing methane increases both CO_2 and C_1 in the vapor stream and decreases the CO_2 capture percentage. Increasing ethane and propane increases both CO_2 and C_1 in the vapor and decreases the CO_2 capture percentage. An experimental bench-scale apparatus verified the simulation results. Exergy loss and heat exchange efficiency have been calculated and equipment sizes and cost estimates are complete. The results confirm the low exergy losses of 39.6 kWh/ kg treated NG for simulation and 44.8 kWh/ kg treated NG for bench-scale apparatus. High heat exchanger efficiency of +90% has been achieved. High pressure and low methane in NG composition results in high CO_2 capture, though the capture efficiency reaches a maximum at intermediate pressure.

512429 Dynamic Intensification of Separation Processes

Abstract:

Process intensification (PI) has captured significant attention as a set of innovative design principles that can dramatically enhance process performance, reducing energy/capital cost and increasing process efficiency[1]. PI technologies usually integrate multiple coupled phenomena into one physical device to optimize mass, heat transfer and reduce equipment size. However, the advantages of PI, including reduced holdups, increased safety and more nimble operation come with some drawbacks and operational barriers. These include a reduction in the number of degrees of freedom (DOF) available for control, as well as a narrower window for optimization compared to conventional process equivalents[2].

In this paper, we will introduce a new dynamic operation paradigm that can be applied to existing intensified equipment to further enhance process performance. In particular, we study the use of periodic operation to manipulate and hijack the competing phenomena present in an intensified system to the benefit of economic indicators such as energy efficiency and yield. We refer to this approach as â€œdynamic process intensification.â€

Motivated by their large energy consumption across the chemical and petrochemical industries, we focus specifically on the dynamic intensification of distillation systems. Using the canonical model of a binary distillation column, we demonstrate theoretically the benefits of dynamic intensification. We then carry out extensive dynamic optimization and simulation studies to confirm the results of the theoretical findings and quantify the magnitude of the economic benefits available in a system of practical relevance.

References

[1] Creative Energy. European Roadmap for Process Intensification. 2007

[2] Schembecker, G.; Tlatlik, S. Process Synthesis for Reactive Separations. Chem. Eng. Process. 2003, 42, 179-189.

Top Dividing Wall Columns in Process Intensification

Manish Bhargava, Roomi Kalita

GTC Technology US LLC., Houston, USA
mbhargava@gtctech.com
rkalita@gtctech.com

Dividing Wall Column (DWC) technology provides refiners with a unique way of improving the efficiency and profitability of the traditional refining techniques. This technology, which has been in use since the 1940s, can be modified for various applications ranging from naphtha/reformate splitters to gas plants and can even be used to transform established technologies like isomerization and naphtha hydrotreating.

In this paper, we discuss the distinct advantages that DWCs can offer to improve and modify traditional gas plant process schemes with emphasis on top dividing wall columns. DWCs are often characterized by the presence of a vertical wall in a regular distillation column. When the distinctive wall is present at the top of the column, the column can be referred to a top dividing wall column.

Top DWCs offer a host of advantages, especially in gas plants which are generally operated under high pressures. Unlike a conventional distillation column, the top of these columns is segregated into two separation zones within the same shell with no intermixing between the two zones. As a result, they can be operated in separate modes and basically behave as two independent columns. Absorption and distillation can be carried out in a single column. This principle forms the basis of GT-LPG MAX® technology.

The concept of Top DWCs can also be implemented in a reformer unit gas plant and a FCC (fluid catalytic cracking) unit gas plant, wherein two or more columns operate in a sequence and under similar operating pressures and temperatures. The DWC will potentially combine the operation of two or more columns, thereby minimizing both capital and energy consumption of the whole sequence.

Top Dividing Wall Columns in LPG Recovery from Off gas

LPG (Liquefied Petroleum Gas) recovery units are an integral part of any refinery. In general, in LPG unit two or more columns remove the C_1-C_2 components along with other non-condensable gases, while concentrating the C_3-C_4 components to generate LPG. Figure 1 shows a standard design consisting of a deethanizer column followed by a depropanizer column. The deethanizer column operates at a high pressure, wherein the C_1-C_2 components are removed as the top product. The C_3 rich (or C_3-C_4) stream is removed in the next column. The recovery of LPG in this design is generally poor due to the high operating pressures and propane losses in the off-gas. In some processes, these losses are mitigated by using refrigeration in the overhead system. This increases the capital cost of the process.

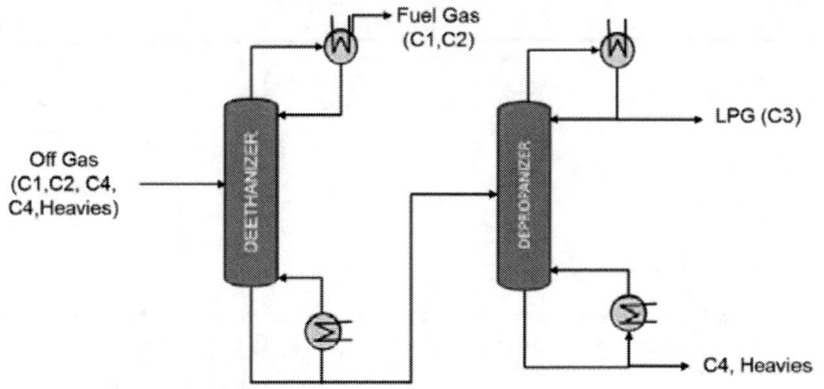

Figure 1: Conventional LPG Recovery Unit

These problems can be rectified by using a top dividing wall column. Doing so allows the top section of the column to be operated under two different unit operations, i.e., absorption and distillation. The section of the column where the feed enters operates under reboiler absorption. The other section of the column separates a light liquid product and a heavy bottom product using regular distillation. This arrangement offers the following advantages.

First, use of absorption allows the column to be operated at a lower pressure. The absorbing solvent captures the valuable C_3-C_4 components in the off-gas and moves them towards the bottom of the column. These components are then concentrated on the other side of the DWC using distillation. For feeds which contain a good amount of C_5 and heavier components, a portion of the heavy bottom product can be used as the absorption medium (Figure 2a). Using the bottoms is highly beneficial when the feed contains at least 30-40 wt% of C_5s. Table 1 gives a sample GT-LPG MAX® feed (Bhargava et al., 2015). If the C_5 concentration in the feed is low, the bottoms solvent can be supplemented by an additional heavy oil solvent from nearby processes.

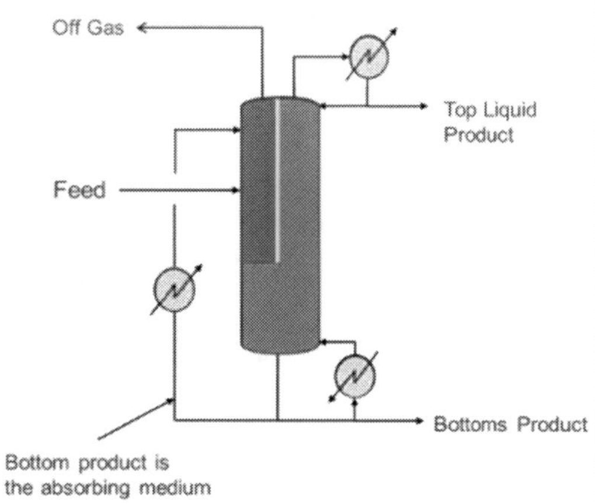

Figure 2a: GT-LPG Max® process scheme; Figure 2b: Inside view of a top dividing wall column

Table 1: Sample Feed Composition for GT-LPG MAX® Design

Product Class	Feed Composition		Product Specifications
	Components	wt.%	
Fuel Gas	Hydrogen	0.0	Minimize Propane
	Methane	0.3	
	Ethylene	0.0	
	Ethane	3.0	
	H_2S	0.5	
	CO_2	0.0	
LPG Product	Propene	0.1	Propane > 90.0 wt.%
	Propane	11.7	Ethane <1.0 wt.%
i-C_4 Product	I-Butane	8.3	Propane < 3.0 wt.% n-C_4 < 4.0 wt.%
n-C_4 Product	n-Butane	19.6	i-C_5 < 3.0 wt.% i-C_4 < 5.3 wt.%
Debutanizer Bottoms	C_5 and heavier	56.6	

Due to the low column operating pressure, water-cooled partial condensation is used in the off-gas side of the column. This is highly economical as compared to using refigeration to mitigate C_3-C_4 losses.

Additionally, this DWC design sees a reduction in energy consumption by about 10-30% as compared to traditional columns with better product recoveries (Table 2).

Table 2: Project Economics of GT-LPG MAX® Design

Variables	Units	Conventional Design		GT-LPG MAX® Design
		Deethanizer	Depropanizer	
Operating Pressure	MPaG	3.1	2.1	2.1
Reboiler Duty	MMKcal/hr	5.5		5.0
Propane Recovery	%	55.0		97.0

Top Dividing Wall Columns in Reformer Unit Gas Plants

A typical reformer unit gas plant usually has a depentanizer column, followed by a debutanizer and a deethanizer. In Figure 3, the depentanizer column removes the C_5- cut from the reformer bottoms as the top product. C_6 and heavier cuts are removed as the bottom product. In the next column, the C_5s are concentrated in the bottom of the column. The C_4 and lighter cut is then processed in a deethanizer column, which removes a LPG cut at the bottom.

The three columns in this sequence operate at high pressures. Additionally, the debutanizer and deethanizer columns see partial overhead condensation. Due to these, a major portion of the LPG components (C_3-C_4) end up in the off-gas. These components can be recovered by use of a dividing wall column configuration.

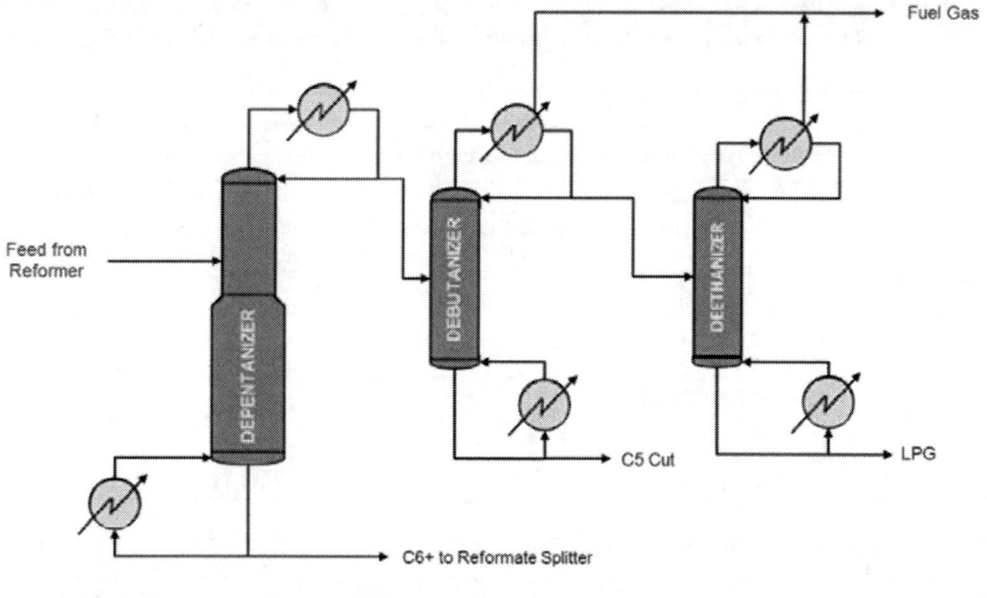

Figure 3: Conventional Reformer Unit Gas Plant Configuration

As shown in Figure 4, the operation of the deethanizer can be squeezed into the bigger depentanizer column by means of a dividing wall. The column has two distinct zones within, a pre-fractionation zone and a main fractionation zone. The pre-fractionation zone sees a combination of absorption and distillation. The lights are separated on this side as the off-gas, with the absorption medium reducing the liquid losses. On the main fractionation side, the C_5 and lighter components are removed as the other top product. Similar to a GT-LPG MAX® design, a portion of the heavy bottoms (mainly C_6-C_7) is used to strip the gas. The GT-DWC® dependanizer column is operated at a lower pressure as compared to the conventional column.

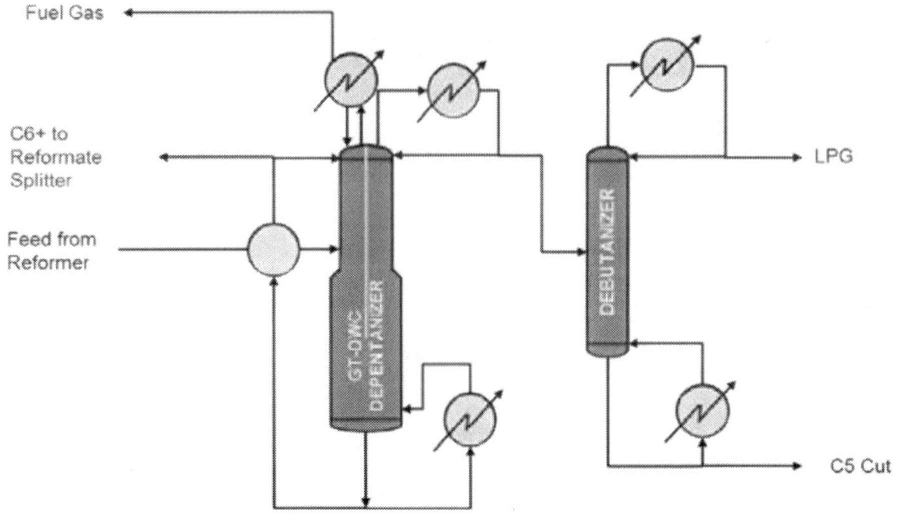

Figure 4: Reformer Gas Plant Configuration with GT-DWC® Technology

Since the majority of the lighter components is removed in the first column, the subsequent debutanizer column operates with a total overhead condenser. This column removes the LPG cut and the C_4 cut. Table 3 compares the performance of the two configurations. For the same heating duties, the DWC configuration generates better product recovery while minimizing the amount of equipment needed. The DWC configuration has a lower total installed cost over the conventional configuration.

Table 3: Performance of GT-DWC® vs Conventional Design

Parameters	Units	Conventional Design			GT-DWC® Design	
Columns		Depentanizer	Debutanizer	Deethanizer	Depentanizer	Debutanizer
No. of Trays	-	40	33	27	40	40
Operating Pressure	MPaG	0.9	1.1	3.1	0.5	1.1
Total Heating Duty	MMkcal/hr	37.1	2.2	0.5	29.0	2.7
Total Installed Cost	MMUSD	10.7			8.5	
Feed	tonnes/hr	386.8			386.8	
C6+ Cut	tonnes/hr	365.5			365.5	
C5- Cut	tonnes/hr	8.1			8.2	
Off Gas	tonnes/hr	5.8			1.7	
LPG	tonnes/hr	7.4			11.4	

Top Dividing Wall Columns in FCC Unit Gas Plants

Figure 5 shows a traditional FCC unit gas plant. The process scheme primarily consists of absorber/stripper columns, where the gases and the liquid in the feed are separated. Unstabilized naphtha (supplemented with a portion of stabilized naphtha bottoms) prevents the C_3-C_4 losses from the absorber column. The remaining C_3-C_4 are extracted by using lean oil in the sponge absorber column. Sour fuel gas and rich oil get separated at the top and bottom. The liquid product from the stripper column is processed in a debutanizer column. Sour LPG is extracted as the top product, while stabilized naphtha is recovered at the bottom.

The conventional design consists of three stages for the separation, the absorbing and stripping section followed by distillation in the main debutanizer column. Using DWC technology, the three stages can be integrated into a single column (Figure 6).

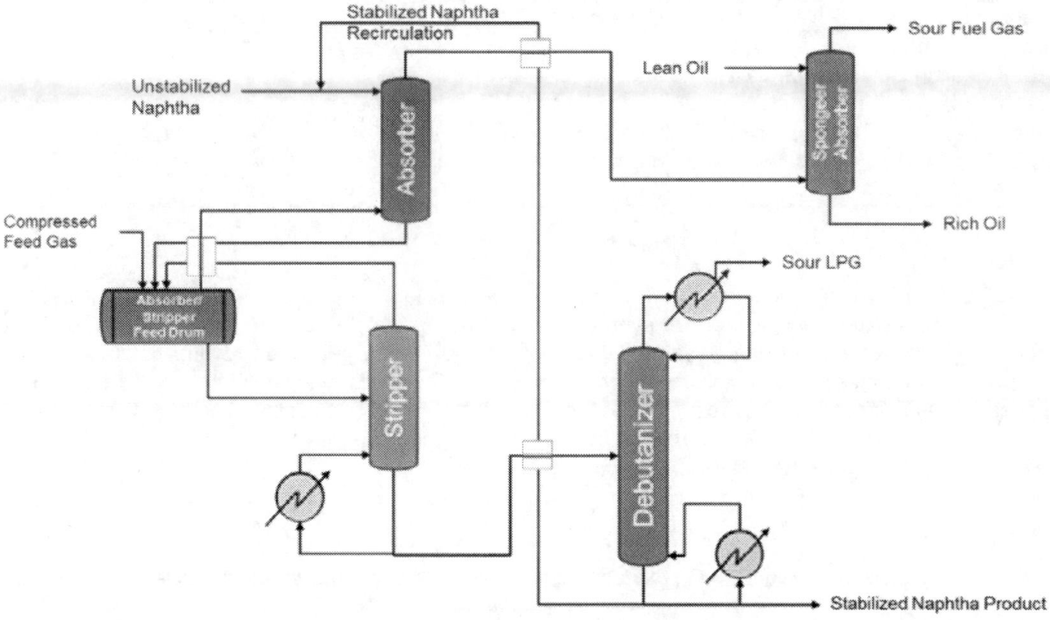

Figure 5: Typical FCC Unit Gas Plant Configuration

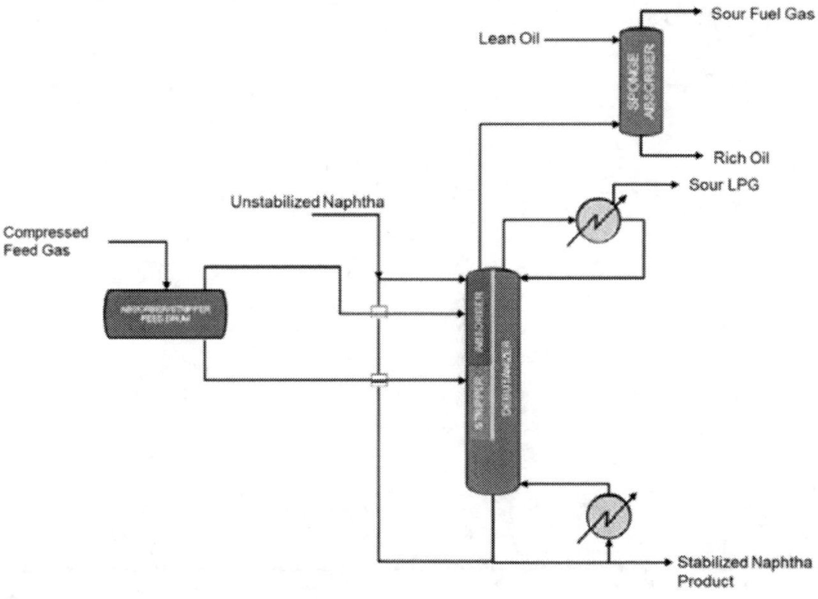

Figure 6: FCC Unit Gas Plant Configuration with GT-DWC® Design

Due to the segregation of the top halves, the two sides of the column behave as individual columns with independent overhead systems and controls. The parallel zones provide similar stages for separation in a single column. A portion of the bottoms product is circulated as the absorption medium in the section where the feed enters. This operation is similar to that of the absorber in the conventional design. The liquid portion of the feed enters below the absorption section. The C_3-C_4 components move down the wall aided by absorption. Once across the wall, these components are concentrated as the top product by rectification and stripping. The stabilized naphtha is removed as the bottom product. Table 4 gives a comparison between the two designs.

Table 4: Comparison between Conventional Design and GT-DWC® Design

Parameters	Units	Conventional Design	GT-DWC® Design
No. of Columns	-	4	2
Total Heating Duty	MMKcal/hr	9.9	7.9
Utility Savings	%	-	20.0
Feed rate	tonnes/hr	55.3	55.3
LPG Recovery	Vol %	85.1	85.1
C4 Recovery	Vol %	98.8	99.6

Conclusion

Gas Plants are a vital component of any refinery. These columns have always been operated using well-established distillation techniques. However, using the Dividing Wall Column concept, primarily top DWCs can generate a leaner and efficient system. DWCs not only lower the equipment count (columns, pumps, and reboilers' et al.), these columns also lower the energy consumption of any configuration. On occasions, these novel distillation columns can even provide better and sometimes new products as compared to the conventional columns. Integrating DWCs in gas plants can virtually change the way the refining industry improves the profitability of their gas plants in the coming years.

References

Bhargava M., Nelson C., Gentry J.C., Siddamshetti, V., 2015, Improved Distillation Efficiency, Hydrocarbon Processing (January)

512585 Use of Dividing Wall Technology to Increase Throughput and Improve Product Specifications in Two-Diameter Columns

Abstract:

Two-diameter section columns are quite common in the refining industry. Standard examples include naphtha splitter columns, stabilizer columns or depropanizer columns. These columns are encountered in applications wherein the process sees different vapor-liquid flow within the top and bottom section of the columns. In such applications, two-diameter section columns are favored by refiners over a single diameter column due to savings in fabrication of such columns. However, for any capacity increase or improving the process separation, this smaller section acts as a bottleneck due to the inability to handle increased vapor-liquid loads. Use of Dividing Wall Column (DWC) technology will allow the refiner to circumnavigate the problem. DWC allows the optimization of the lower diameter section to achieve better separation as well handle the increased capacity from the original design.. Two case studies based on actual installations will be presented:

- A conventional two-column fractionation sequence is revamped to a single two-diameter DWC column to produce the same product specifications as the original design. The second column in the sequence is idled following the revamp.
- An abandoned two diameter column is revamped for maximizing IC5 Recovery.

Authors:

- Manish Bhargava, mbhargava@gtctech.com
- Roomi Kalita, rkalita@gtctech.com

GTC Technology

900 Threadneedle St., Suite 800

Houston, TX 77079

281-597-4800

513423 Process Intensification of Distillation Using a Microwick Technology Platform

Abstract:

Process intensification of distillation has been advanced using a microwick technology platform that has been demonstrated for multiple gas-liquid processes, including partial condensation, absorption, stripping, and distillation. Mass transfer processes are intensified by constraining the liquid phase within thin wicking structures, which for distillation dramatically reduces the height of an equivalent theoretical plate (HETP). Multiple separations have been demonstrated with microchannel distillation (MCD) devices, including fractionation of JP-8, a military diesel fuel, purification of n-octane, separation of propane and propylene, and isotopic enrichment of methane. Results have yielded HETPs as low as 0.33 cm and up to 60 separation stages were accompished in a single device with an active length of 25 cm. Dimensional analysis of the data from the variety of separations provide insights into the performance and scaleability of the technology. Computational fluid dynamics models of the processes indicate the potential for an additional factor of 10 reduction in HETP. Comparisons are made to other microfluidic distillation technologies as well as to commercial advanced structured packings.

513707 Improving Plant Economics By Upgrading Low Value Hydrocarbon Stream By Enrichment of C_{9+} Compounds Using Divided Wall Distillation Columns

Abstract:

The shift towards lighter feedstockâ€™s for Poly Olefins production, due to lower feedstock price and higher profit margin, has resulted in lower production of heavier aromatic fractions and in particular the C_{9+} fractions. As a result, over the last 5 years the supply of C_{9+} fractions has not catch up with increase in demand. Further, intense competition between petrochemical complexes has forced companies to retrofit plants for comprehensive utilization of all streams in the complexes for better operating margin. Among other things that are considered for up-gradation, enrichment of heavier hydrocarbons from the Pygas stream has been pursued intensely. In this work, the process intensification in C_{9+} stream enrichment from Pygas using non-conventional divided wall column structures is examined. The study is a first of kind to demonstrate the separation of bulk chemicals using divided wall columns. Further more, the studies indicated that by replacing conventional column sequence with the divided wall column structure, for a given product quality, a savings in the separation stages by 18%, energy reduction by around 32% can be realized.

513072 Intensified-By-Design (IbD): Creating a Platform for Facilitating Process Intensification in Solids Handling Applications

Abstract:

Since its inception, the definition of process intensification (PI) has been broadened from apparatus-centric â€˜miniaturisationâ€™ to the development of processing techniques that offer a drastic reduction in spatial- and energetic-requirements as well as in waste production. With the resulting enhancements in mixing, heat/mass transfer and all other benefits due to PI, many processes involving gas and liquid systems are becoming more proficient in performance and efficiency. On the other hand, the lack of focus in PI for processes involving solids handling has been attributed to the perception of fouling or blockage potentials in micro channels which have often been the technology of choice for miniaturisation. To address this, a clear understanding of industrially-relevant solids handling processes must be undertaken for proper equipment design, thereby remediating the aforementioned hindrances and hence, help facilitate the implementation of PI. These processes have been classified as precipitation/crystallisation, separation, granulation, mixing, particle coating, milling/grinding, catalytic reactions, particle classification, drying, thermal processing and bioprocessing.

In this presentation, five relevant powders- and solids-handling PI technologies are highlighted: (1) The Spinning Disk Reactor (SDR) applies a high centrifugal acceleration to liquids flowing on its surface, forming an extremely thin wavy film with a high surface area to volume ratio. This results in a rapid micro-mixing, creating high supersaturation levels, suitable for precipitation. (2) The Oscillatory Baffled Reactor (OBR) involves a fluidic oscillation in tubes equipped with baffles, resulting in enhanced mixing and heat/mass transfer. This technology has been applied for continuous crystallisation, particle suspension and catalytic reactions. (3) The Taylor-Couette reactors (TCRs) are composed of two differentially rotating concentric cylinders which are separated by a small gap. Flow within the annular space of the Taylor-Couette device experiences a sequence of hydrodynamic instabilities, from laminar Taylor vortex to turbulent regimes, as the rotational speed of the cylinder increased. This ï¬‚ow dynamics is usually characterised by the rotational Reynolds number or Taylor number. The vortex structure in various regimes of hydrodynamic instabilities could be exploited for various potential applications, including particles classification, crystallisation and granulation. (4) Microfluidics deal with the behaviour, precise control and manipulation of fluids that are geometrically constrained to a small scale, typically sub-millimetre. Micro-fluidized beds, for example, utilise the excellent mass and heat transfer characteristics in many applications e.g. kinetics measurement in solid catalysed reactions, detection of infectious bacteria, etc. (5) Twin screw granulator (TSG) is set to transform pharmaceutical processing by offering continuous granulation of drug formulations (currently carried out in batch), thus complying with Quality-by-Design concept, which highlights an automated real-time quality monitoring and control in tablet production.

For these and other solids-related PI technologies investigated in the Intensified-by-Design (IbD) project (funded by the EU Horizon 2020), detailed processing, control and fouling data are compiled in a cloud-based platform. The platform includes built-in design modules, which output datasets for users to facilitate reactor design and assembly, as well as a holistic process optimisation i.e. upstream and downstream intensified unit operations, solids handling capability, cleaning methods and the expected economic and environmental quantitative impacts.

519833 Rapid, In-line Characterization to Support Process Intensification Concepts

Abstract:

Products consisting of processed particulate solids are utilized at some point in the majority of many commodity industries, including the food, pharmaceutical, agricultural and chemical industries. Careful monitoring and control of the manufacturing process is crucial to maintaining high product quality, especially in situations in which process intensification is employed because individual operations often require more stringent specifications and also each operation has greater impact on product properties (i.e. intensification of changes to the product increases the feedback/controls challenge). Particulate solid materials are often much more difficult to handle and feed than liquids or gases, which also exacerbates materials processing difficulties. This talk presents methods to assess material handling and feeding properties of particulate solids, including particle size/shape distributions, density, cohesion, and internal angle of friction (all of which are metrics of flowability). 3D laser scanners and a 2D digital camera are used to automatically measure particle size/shape distributions on a moving conveyor belt, and a custom hopper with adaptable geometry is used to measured density, cohesion, and internal angle of friction. Flow properties are determined by continually adjusting the size of the outlet in a plane-flow hopper to maintain a target fill level in the hopper. The angle of the hopper walls can also be adjusted to tune the geometry to different materials. Test results, including wall pressure measurements, are compared to simulations and indicate that these methods show promise for in-line quantitative estimation of the flow properties of compressible and anisotropic materials.

512700 Hybrid Heat Pipe Screw Dryer: A Novel, Continuous and Highly Energy-Efficient Drying Technology

Abstract:

This work investigated a novel type of screw conveyor dryer (SCD) equipped with a passive heating provided by an annular heat pipe, termed "Heat Pipe Screw Dryer" (HPSD). The device was designed as a compact and continuous drying equipment with an energy efficient operation as its core principle. The heat pipe was firstly characterised, resulting in the lowest thermal resistance of 5.6 x 10^{-3} °C/W at certain conditions: 200°C setting temperature, 11% filling ratio and in (laid flat) horizontal orientation. These conditions therefore, were employed in the drying of a ceramic raw material slurry (initial moisture content: 33 wt. %, wet basis). Moisture reduction of the material increased with increasing temperature and residence time. Approximately 46% moisture reduction (corresponding to 18 wt. % final moisture content) was achieved at 200°C and 260s conditions with the resulting energy efficiency of 2.04 kg/kWh, which places the HPSD as more energy efficient than conventional dryers e.g. fluidised beds, rotary and spray dryers. In addition, thermal analyses showed that the HPSD performed better than a current steam jacketed-SCD, by ~35% higher in the overall heat transfer coefficients. These were attributed to the excellence of the annular heat pipe as a thermal conductor as well as the negation of any pumping requirement of heating fluid. On the other hand, solids buildup (fouling) also increased with moisture reduction, from 3 to 62% for 10 to 46% moisture reductions, respectively. Applying a non-stick "Xylan" coating on the screw conveyor resulted in a ~10% fouling remediation. From experimental data, it was concluded that the fouling could be improved significantly if the entire inner barrel was coated. Reducing the barrel clearance and periodic wash-ups were among other fouling remediation strategies. The HPSD as a multifunctional technology, combining both drying and heating, is a worthy technology to be further investigated, developed and applied, creating a more energy-efficient drying processing.

513193 Process Intensification and Cost Reduction for in-Plant Biomass Preprocessing Equipment

Abstract:

Process intensification and cost reduction for in-plant biomass preprocessing equipment

Michael Blaise[*], *Nicholas, R. Schwartz, Paul E. Yelvington* Mainstream Engineering Corporation, Rockledge, FL [*]Corresponding Author: mblaise@mainstream-engr.com Abstract

With growing concern about the sustainability and environmental impact of fossil fuels, interest in fuels derived from biomass feedstocks (e.g., agricultural residues, forestry residues, energy crops, and organics in mixed solids wastes) has risen steadily. Designing low-cost feedstock preprocessing systems that can handle a variety of materials is crucial to ensuring efficient and economic plant operation. Conventional preprocessing and solids handling consists of size reduction (i.e., comminution), drying, conveyance, and an inerting system to purge and pressurize the feedstock prior to introduction into a substantially oxygen-free conversion reactor such as a fluidized-bed fast-pyrolysis reactor. In this work, process intensification of preprocessing equipment for thermochemical biomass conversion was designed to take advantage of existing hot flue gas and eliminate the need for separate energy-intensive drying or purging steps. This approach combines drying, knife milling with in-line sharpening, conveyance, and inerting functions. This process-intensified approach is referred to as the integrated DCIMS (drying, conveyance, inerting, milling-sharpening) in-plant feedstock preprocessing. The DCIMS approach eliminated the need for a rotary drum dryer and second conveyor compared to the baseline conventional approach. A better than 5% additional reduction of moisture content was achieved when heating the knife mill with hot flue gas compared to an unheated baseline. Synergy between drying and milling led to reduced heat duty and reduced mill shaft power through combining these processes. The DCIMS approach demonstrated a more uniform (and smaller) milled product without increasing the energy required for milling. Experiments showed that the final moisture content for pine feedstock was 9% (a 15% reduction from a 24% starting moisture), for sugar cane bagasse was 8% (a 20% reduction from a 28% starting moisture), and for mixed waste was 5% (a 25% reduction from a 30% starting moisture). Internal integrated sharpening was also demonstrated successfully. The quality of the blade improved significantlyâ €"feedstock product was more uniform, and the power required for milling decreased by 17%. The estimated operating cost savings from the internal sharpening system was 4%. Heated pneumatic conveyance was also shown to effectively perform biomass drying. A custom motive nozzle for an eductor allowed for low-pressure flue gas to be used directly in the pneumatic conveyance of the feedstock. Integrated DCIMS feedstock preprocessing reduced the operating costs by 37% to $27/dry ton at a 1,000 ton/day scale, based on a technoeconomic analysis (TEA) of the process.

Keywords: Fast pyrolysis, process intensification, preprocessing, biomass, bio-oil

512932 Overcoming Heat Transfer Bottlenecks in Non-Equilibrium Chemical Processes through Autothermal Operation

Abstract:

Overcoming Heat Transfer Bottlenecks in Non-Equilibrium Chemical Processes through Autothermal Operation

Robert C. Brown[1, 2] and Chad Peterson[2]

[1]Bioeconomy Institute

[2] Department of Mechanical Engineering

Iowa State University

Innovations in Process Intensification Session

Topical C: Process Intensification

AIChE Spring Meeting, Orlando, FL

April 22-26, 2018

Abstract

Heat transfer is the bottleneck in many chemical processes, limiting throughput for a reactor even when chemical reaction and mass transfer are very fast. Efforts to enhance heat transfer can add significantly to equipment costs. Furthermore, with heat transfer rate limiting, reactor capacity only scales as the square of reactor diameter, constraining throughput as reactors are built larger and increasing capital costs faster than expected from economies of scale principles.

Autothermal operation, which balances the energy demand of endothermic chemical reactions with energy supply from exothermic chemical reactions, can overcome heat transfer bottlenecks. Although autothermal operation has been previously demonstrated for *equilibrium* chemical processes (prominent examples include methane steam reformers and biomass/coal gasifiers), application to *non-equilibrium* processes has been little explored. Identifying complementary endothermic and exothermic reactions that can occur simultaneously while preserving the same products of an externally heated (or cooled) chemical process is not trivial. However, the gains in *process intensification* would be substantial, increasing throughput and reducing capital and operating costs of a chemical process.

We have experimentally demonstrated the principle of non-equilibrium autothermal operation for the fast pyrolysis of biomass. Autothermal pyrolysis was achieved by admitting a small amount of air into a fluidized bed pyrolyzer, which partially oxidizes biomass reactant or pyrolysis products to provide the enthalpy of pyrolysis without substantially affecting the desired pyrolysis products. Decreases in non-condensable gases such as carbon monoxide and methane during autothermal pyrolysis would suggest their oxidation is responsible for a significant fraction of the exothermic energy released during this process. However, autoignition temperatures for the non-condensable gases in pyrolysis products is higher than typically occurs in fast pyrolysis, suggesting other exothermic chemical reactions provide the energy to support autothermal pyrolysis.

We are developing a chemical kinetic model of autothermal pyrolysis to better understand the source of energy for the process. The model indicates that non-condensable gases are not substantially oxidized during autothermal pyrolysis. Instead, precursors of these gases along with other light oxygenated compounds are oxidizing and providing the enthalpy for pyrolysis. The model is also investigating solid-gas reactions to investigate the contribution of lignin and char to the energy required for autothermal pyrolysis.

512574 Reverse Flow Reactors: Process Intensification for High Temperature Chemistries

Abstract:

Global chemicals demand is expected to surpass 200 million metric tons by the end of this decade, with a growth pace faster than the global GDP. The precursors to these chemicals are often manufactured using high temperature endothermic reactions such as reforming and pyrolysis. Substantial energy is consumed to heat feed streams to reaction temperatures, and ineffectively captured as product streams are further processed. Additionally, the significant capital, physical and carbon footprint requirements make these processes ideal for a process intensification (PI) play.

We describe a reverse flow reactor (RFR) approach that efficiently integrates heat-transfer, chemical reactions and cyclical in-situ de-coking by alternating flows between reaction (endothermic) and reheat (exothermic) steps in a fixed bed reactor. The use of high surface area, thermally robust ceramic internals allow operating temperatures up to 1500°C enabling higher conversion and short residence times (higher selectivity) long sought out for optimal performance.

This presentation will provide a brief introduction to RFR, discuss the key reactor features and exemplify a few applications. Finally, the need for a multi-disciplinary approach involving interplay of modeling, materials development, and experimental demonstration is advocated for the advancement of new PI technologies.

517124 Building with Blocks

Abstract:

Chemical processing facilities traditionally use a â€œstick buildâ€ approach for expansions, retrofits and new plants. This involves field installation and erection of all equipment, piping and components. Costs for this strategy are well understood which perpetuates its continued preference. However, this may not always be the most cost effective approach or the most appropriate strategy. In some instances Modular Manufacturing offers a lower cost alternative to conventional methods. Modules can be thought of as the building blocks from which processes can be built. In small scale manufacturing these modules can be designed with considerable flexibility and bolted on to for enhanced capability. In some cases modules can be assembled as a faster and lower cost way to build a large facility, particularly when local skilled resources are a limitation.

This presentation will highlight beneficial applications of modular manufacturing utilized in a centralized facility. The presentation will also include identified knowledge gaps that limit a broader use of the technology. More cost information and understanding is needed to know when the best approach is to build with blocks or sticks.

517126 Critical Factors and Criteria for the Development and Deployment of Modular Chemical Processes

Abstract:

The allure of implementing modular chemical processes is the possibility of achieving significant cost savings compared with traditional individually-engineered and stick-built processes. In theory a factory-built, modular process could have a lower construction cost than a stick-built process; however, factory-built processes must be constructed of units that are a size that can economically be shipped and connected in the field. For processes above a certain size, this would necessitate the traditional engineering scaling factors being broken via advances in technology, manufacturing, and engineering design approaches. While the concept of producing modular processes is appealing, overcoming these traditional scaling factors is challenging because current supply chain and manufacturing systems are geared towards the stick-built approach for chemical process equipment. These cost savings will only be achievable if multiple standard units of the same design are produced. System standardization and minimization of one-time engineering and maintenance costs are essential to minimizing total cost of ownership of a modular system.

Considering these constraints and factors, purification processes for wastes or feed preparation are low-hanging fruit for the implementation of modular chemical processes for both distributed and centralized applications. Implementation of modular processes for distributed and centralized locations have different primary challenges. For the case of centralized applications ensuring ease of integration with the existing infrastructure, balance of plant, and controls becomes paramount, while for distributed applications it is essential that the process is highly autonomous, robust, and self-sufficient.

This talk will summarize and examine the critical factors and criteria for the development and deployment of modular chemical processes. Topics to be discussed will include when to choose modular, system size selection, process controls, standardization, manufacturing, and design for CAPEX and OPEX minimization. A focus will be placed on examining these factors relative to IntraMicron's recently developed Sour Gas Sweetening Process, SourCat^{TM}. The SourCat^{TM} desulfurization process is currently being commercialized by SourGas LLC, a wholly owned subsidiary of IntraMicron. The SourCat^{TM} process will be deployed commercially as distributed modular units in the Oil and Gas Industry. The transition of this process from a modular pilot plant to a modular chemical process is currently the subject of the RAPID program "Modular Catalytic Desulfurization Units for Sour Gas Sweetening."

517136 Continuous Processing of Specialty Chemicals â€“ Opportunities with MCPI

Abstract:

Specialty chemicals and small commodities market is challenged with relatively small throughputs and relatively high production cost. Dissemination of the costs is prerequisite for making business decision for capital investment in new capacity. The new plant must be near to the market and be able to adjust rapidly to market changes. It must also have possibility for flexible production e.g. change of the product must be relatively simple. Fulfillment of these demands is challenging for the process engineering. In order to mitigate these challenges, design of cheaper, intensified modules, are of most importance since the solution may lay in modular plant and/or in process intensified equipment such as flow reactors or dividing wall columns. In order to discover and utilize the opportunities a dissemination of capital expenditure is essential.

AUTHOR INDEX

Alqahtani, Abdullah 17
Baldea, Michael 8
Barrai, Federico 23
Bhandari, Dhaval 23
Bhargava, Manish 9, 15
Blaise, Michael J. 21
Boodhoo, Kamelia 18, 20
Bottenus, Danny 16
Brown, Robert C. 22
Chun, Changmin 23
Dimick, Paul S. 5, 25
Edgar, Thomas F. 8
Eze, Valentine C. 18
Fazlollahi, Farhad 7
Fleming, Brad 9
Hershkowitz, Frank 23
Humble, Paul 16
Kalita, Roomi 9, 15
Kumar Srinivas, Mahesh 17
Lattner, James 23
Luo, Yang 4
Mahoney, Elizabeth G. 23
Mountziaris, T. J. 2
Mustaffar, Ahmad 18, 20
Pajalic, Oleg 26
Peterson, Chad 22
Phan, Anh 18, 20
Powell, Michael R. 16
Ramkumar, Shwetha 6
Sastri, Bhima 1
Schwartz, Nicholas R. 21
Shatto, Don P. 6
Stark, Addison K. 3
Tatarchuk, Bruce 25
Tatarchuk, Bruce J. 5
Tegrotenhuis, Ward E. 16
Theunick, Greg 24
Weiss, Brian M. 23
Westover, Tyler L. 19
Yan, Lingqing 8
Yang, Hongyun 5
Yeh, Norman K. 6
Yelvington, Paul E. 21
Zivkovic, Vladimir 18